D0801250

Ingredients (makes one human):   oxygen 61%, carbon 23%,
hydrogen 10%, nitrogen 2.6%, calcium 1.4%, phosphorus 1.1%,
potassium 0.2%, sulphur 0.2%, sodium 0.1%, chlorine 0.1%
plus magnesium, iron, fluorine, zinc and other trace elements.

First published 2003 AD
This revised edition © Wooden Books 2012 AD
Published by Wooden Books Ltd.
Glastonbury, Somerset

British Library Cataloguing in Publication Data
Tweed, M.
Essential Elements

A CIP catalogue record for this book is
available from the British Library

ISBN 1 904263 58 5

Printed and bound in Shanghai, China
by Shanghai iPrinting Co., Ltd.
100% recycled papers.

ESSENTIAL ELEMENTS

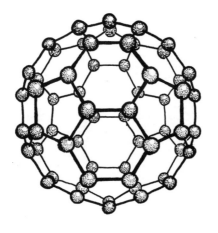

written and illustrated by

Matt Tweed

With love …

Thanks to the Blackaby family for all their support,
& as always my Mum for being great!

"… that which is above is as that which is below, and that which is below
is as that which is above, to perform the miracles of the One Thing."

The Emerald Tablet of Hermes Trismegistos

Contents

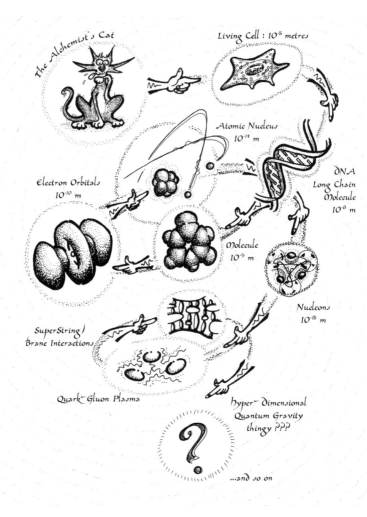

The Alchemist's Cat

Living Cell : 10^{-5} metres

Atomic Nucleus
10^{-14} m

Electron Orbitals
10^{-10} m

DNA
Long Chain
Molecule
10^{-8} m

Molecule
10^{-9} m

Nucleons
10^{-15} m

SuperString/
Brane Interactions

Quark-Gluon Plasma

Hyper-dimensional
Quantum Gravity
thingy ???

...and so on

INTRODUCTION

Pretty much all that we see or touch in our seemingly solid existence is made from squintillions of tiny atoms, each being one of over a hundred unique types of element. Combined together in a myriad different ways, they form the fantastic mosaic that is the visible universe.

If we peer closely at an individual atom, the first astonishment is that it is mainly empty space. Fizzball electrons spin complex webs around a central nucleus, a miniscule point in the middle of a galaxy of whirling energy. Even here we are just scratching the surface—beyond are places where the rules become very strange indeed, where solidity has little meaning and matter comes in waves. Whole sub-atomic families appear, and particles interfere, tunnel, entangle and generally indulge in behaviours which defy common-sense, yet follow their own probabilistic laws. This is a place of high energies and fundamental forces that directly shapes our macroscopic experience.

Zooming in further still, we find that even this tiny kingdom may itself be the knottings of ever more ephemeral wisps on the very edges of our understanding, held together in symmetric patterns that span dimensions and perform a dance of deep mathematics.

All are players in this great game of life, acts of consciousness interlacing and resonating through the all that is.

Most of all I hope you, dear reader, will enjoy this brief journey into the wonderful world of matter. May we use this extraordinary knowledge with wisdom and understanding in the millennia ahead.

EARLY ALCHEMY
a wee bit of magick

The roots of chemistry stretch far back into the dim and distant past, to when our ancestors first prepared coloured earths for painting cave walls and themselves, learned the secrets of fire, and started experimenting with the arcane intricacies of cookery.

The ancient Egyptians knew of seven metals, as well as non-metals such as carbon and sulphur, all easily extracted from natural ores. The art of *Khemia,* supposedly revealed by angels, linked the metals to the seven known planets and assigned them unique qualities (*opposite top left*). Antique Indian treatises speak of the three *gunas,* fire, earth and water. Chinese sages used two more, metal and wood (*opposite top right*).

To the later Greek philosophers all things were made of earth, air, fire, or water (*opposite, lower left*). Naming them *elements,* Aristotle, in the 3rd century BC, added a fifth, *quintessence,* which formed the heavens. Another philosopher, Democritus, proposed that dividing matter over and over again would eventually leave an indivisible *atmos.* Scorned by Aristotle, the *atom* was then largely forgotten for centuries.

With the fall of the Greek empire, investigation of *Al-khemia* moved to Arabia. Books like Al-Razi's 10th century *The Secret of Secrets* and Jabir ibn-Hayyan's *The Sum of Perfection* told of an *elixir of life* that could grant immortality and transmute base metals into gold.

The quest spread to medieval Europe where, in the 13th and 14th centuries, alchemists like Albertus Magnus, Roger Bacon and Nicholas Flamel hoped to find the all-powerful *Gloria Mundi* or *Philosopher's Stone.* Slowly, through experiment, trial, error, intuition and the odd happy accident, they laid out the foundations of a extraordinary body of lore.

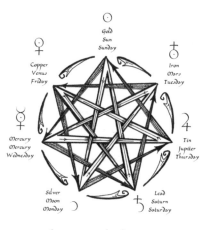

The Seven Metals of Antiquity
The Seven known Planets

Wu-Hsing : the Five-Fold Chinese
Elemental System

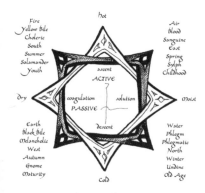

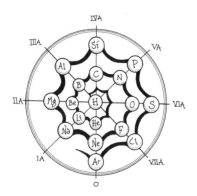

The Four Elements
and corresponding Humours

The Periodic Web of the
First Eighteen Elements

THE AGE OF SCIENCE
alchemy transmutes into chemistry

By the eighteenth century scientists were largely freeing themselves of metaphysical concerns, and early experiments comparing weights and masses showed that many substances assumed to be elemental were in fact *molecules* or compounds of several parts.

In 1789 the first table of twenty-three elements was published by Antoine Lavoisier, soon followed by John Dalton's 1808 inspired forerunner to atomic theory (which was then ignored for fifty years).

As scientific techniques improved, new elements were discovered at a prodigious rate. Noticing how those with similar chemical properties fell into recurring patterns, Dmitri Mendeleev created his famed *Periodic Table of Elements* in 1869, successfully predicting the existence of scandium and germanium. The first hint of stuff smaller than the atom came in 1896, when, unwittingly leaving pitchblende (a uranium ore) on an unexposed photographic plate, Becquerel accidentally discovered radioactivity.

In the early twentieth century Ernest Rutherford's discovery of the surprisingly empty space around the atomic nucleus, the unveiling of the electron orbitals and Albert Einstein's theory that matter and energy were the same thing led Max Planck, Erwin Schrödinger, Niels Bohr and others to the curiously wavy world of quantum mechanics. In 1932 the atom was split for the first time and for the rest of the century scientists explored the symmetries of the sub-atomic realms. Huge smashers hurled atoms together to synthesise new heavy elements, at other times breaking them apart to reveal whole families of exotic particles.

The universe was made of very strange things indeed.

95% Copper with 5% Tin heated to 1100 °C in charcoal furnace

Melt wax out

Clay mould

Wax Model

Molten bronze poured into mould

Break the mould

Polish & Finish

Bronze was one of the first alloys, often cast using the lost wax process

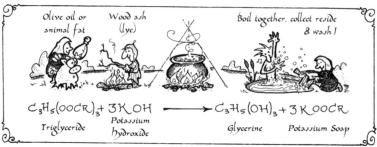

Olive oil or animal fat

Wood ash (lye)

Boil together, collect reside & wash !

$$C_3H_5(OOCR)_3 + 3KOH \longrightarrow C_3H_5(OH)_3 + 3KOOCR$$

Triglyceride   Potassium hydroxide   Glycerine   Potassium Soap

Soaps were perhaps discovered from fat falling into the dampened ashes of a fire

Humphrey Davy
Group 1 & II Elements

Dimitry Mandeleev
Periodic Table

Ramsay & Travers
Noble Gases

Marie & Pierre Curie
Radium & Polonium

Berkeley & Dubna scientists
Transuranium Elements

Some of the many involved in the discovery of the elements

Inside the Atom
proton, neutron and electron

Atoms consist of a small central *nucleus* orbited by one or more whirling *electrons*. Two visualizations of an atom are shown opposite. The nucleus, a mere hundred billionth of a millimeter across, contains two similarly sized particles, *protons* and *neutrons*. Each proton has a single positive electric charge and a corresponding negatively charged electron. It is the proton count which gives an element its name and *atomic number*, or position in the periodic table (*see pages 56–57*). Neutrons have no charge.

Although the number of protons and electrons are fixed in every element, the number of neutrons can vary, giving *isotopes*, which react the same chemically yet can behave quite differently at a nuclear level.

Electrons weigh almost two thousand times less than protons and neutrons. Repelling other negative electrons they are attracted to the oppositely charged positive protons but ignore the chargeless neutrons. Balancing all these forces, electrons team up in pairs and whizz about the nucleus in *orbitals* grouped into orbital *sets*, three dimensional patterns that get increasingly complex in larger atoms. Orbitals fill up in a specific order (*lower opposite*).

Amazingly, atoms are almost entirely empty space. An electron orbiting a nucleus may be visualised as a cat swinging a bumblebee on the end of a half-mile long piece of elastic.

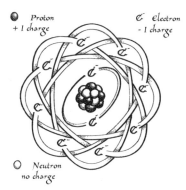

Proton
+ 1 charge

℮ Electron
- 1 charge

○ Neutron
no charge

A classical planetary picture of a neon atom:
a central nucleus of 10 neutrons and 10
protons surrounded by a whirl of 10 electrons,
two in an inner orbit, the remainder in an outer.

A quantum mechanical view of the same atom:
here each lobe represents the probability of finding
an electron in a particular place as given by the
Schrödinger wave equation.

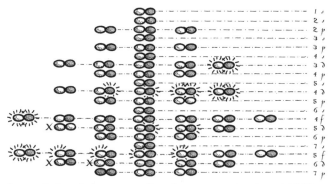

	1 s
	2 s
	2 p
	3 s
	3 p
	4 s
	3 d
	4 p
	5 s
	4 d
	5 p
	6 s
	4 f
	5 d
	6 p
	7 s
	5 f
	6 d
	7 p

Electron orbital sets build in sequence from the innermost 1 s. Each row half-fills
with electrons (white blobs) before completing as oppositely spinning pairs (black blobs).
Glows around a blob indicate that one or two electrons skip to or from other orbitals,
breaking the rigid pattern. Gold, silver and copper are amongst those that share this
quirk. X marks the d orbitals that try to fill before the row above gets going.

7

Periodic Tables
elemental ordering

Every element has its own place in the periodic table, and there are several versions of the table that emphasise different features.

Professor Benfey's spiral (*below*) develops by atomic number (the count of protons in each element) and shows *groups* with the same pattern of outer electrons (and hence corresponding properties) radiating like spokes from a hydrogen hub. As the different orbitals fill, *blocks* of related elements form outcrops.

In a contrasting scheme, Dr Stowe's table (*opposite top*) displays the physical ordering of the intricate orbital sets of electron *shells*, with the innermost at the top, using each element's unique set of *quantum numbers*.

A modern version of Mendeleyev's original table (*lower opposite*) puts groups in vertical columns with horizontal *periods* of orbital sets. Elements are arranged by atomic number, reading left to right, top to bottom.

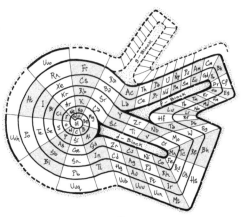

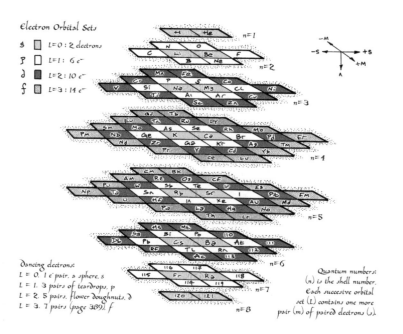

Above: The Atomic Shells, with the shaded orbital sets of which they are comprised.
Below: The modern periodic table, which is read left to right across the open spaces and shows increasing order of atomic number (see pages 56–57).

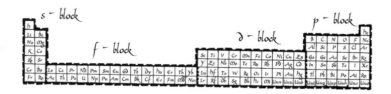

A Burning Question
chemical conflagrations

Most things around us are *compounds*, or combinations of various elements. To bond, atoms rearrange their outer, highest energy *valence* electrons, keeping their full interiors safely out of the way.

Opposite top is an *exothermic* reaction, which produces heat. After initially shaking apart the gas molecules with the heat of a flame, new water bonds quickly form which lock up less energy than the original gas bonds. The released energy keeps the reaction going and the gases explode, frantically shuffling electrons between each other. Lower opposite we see an *endothermic* reaction that takes place when plants *photosynthesise*. Here the sums go the other way round and heat needs to be absorbed, in this case from sunlight. The products therefore have more energy than the reactants, glucose storing the energy. When the reaction moves in the opposite direction, it is known as *respiration*.

Matter itself can exist in several different states or *phases (below)*. *Solids* pack atoms closely in rigid arrangements. Heating vibrates the atoms, shaking the structures apart to form *liquids* that can flow and change shape. Heating further weakens even these loose bonds and the atoms scatter at high speed in all directions as *gases*. At yet higher temperatures some of the electrons are knocked off the atoms to create an electrically charged, ionised *plasma*, like that found in the superhot corona of the Sun.

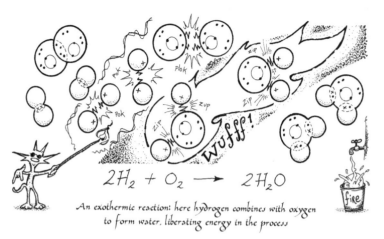

$$2H_2 + O_2 \longrightarrow 2H_2O$$

An exothermic reaction: here hydrogen combines with oxygen
to form water, liberating energy in the process

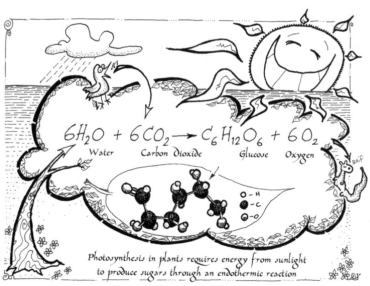

$$6H_2O + 6CO_2 \longrightarrow C_6H_{12}O_6 + 6O_2$$

Water   Carbon dioxide   Glucose   Oxygen

Photosynthesis in plants requires energy from sunlight
to produce sugars through an endothermic reaction

11

BONDING
atomic stickiness

Molecules are formed as atoms' outer electrons share dances. Losing or gaining electrons causes atoms to become positively or negatively electrically charged *ions*. Most elements are either metals, which are *electropositive*, losing electrons to form *cations*, or non-metals, which are *electronegative*, grabbing electrons to form *anions*.

An *ionic bond* occurs when a positive cation lends electrons to a negative anion to give them both full stable outer orbitals like the nearest noble gas (*opposite top left*). Though tough and brittle with high melting points, many ionic compounds dissolve in water.

Non-metals combine using *covalent bonds*, which shuffle and share outer electrons into pairs, again filling up any empty orbitals (*opposite top right*). The attraction felt by electrons for nuclei, outweighing their mutual repulsions, holds the resulting molecule together.

In *metallic bonds* electrons float away from their nuclei, *dissociating* into a 'sea' around a lattice of positive ions (*opposite*). The conductivity and shininess of metals is the direct result of these mobile electrons, and their strength and high melting points result from the lovestruck relationship between the ions and their mates.

Hydrogen attached to a non-metal pushes against unbonded *lone pair* electrons creating a very slight charge difference across the molecule. If another electronegative atom is nearby, a weak *hydrogen bond*, vital in water and DNA, appears between them (*opposite lower left*).

With asymmetrical motions of electrons causing instantaneous small *Van der Waals forces* between atoms, and overlapping orbitals smearing π-bonds (*opposite lower right*), atomic glues come in many forms.

Li⁺   Fl⁻

Lithium Flouride Ionically Bonding

O   C   O

Covalent Bonding in Carbon Dioxide

Pseudo-Electron Density Map of Crystalline Lithium Flouride
(Nuclei centres are 200.9 pico-metres apart)

The Noble Ship of Current Sails the Metallically Bonded Dissociated Sea of Electrons

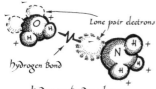

Lone pair electrons

hydrogen Bond

Hydrogen Bonding between
Water (H_2O) and
Ammonia (NH_3)

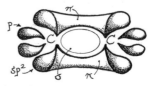

π-Bonds formed from overlapping orbitals
alongside a covalent σ-bond create the
Carbon double bond in ethylene

13

CRYSTALS
building bigger

Crystals are simple repeated patterns of unit cells. Like apples on a market stall, atoms and molecules can grow into large three-dimensional solids, the pieces positioned to give the best balance between attractive and repulsive forces. Ordered in their zillions, these tiny blocks build into many of the solid substances of our world, bridging the vast difference of scale between molecules, minerals and mountains.

There are seven crystalline systems based on tessellating geometries which, when combined with the four basic unit cell types, give the fourteen *Bravais Lattices* (*opposite*). Variations in temperature or pressure may change one crystal structure into another more comfortable and efficient arrangement. Sulphur, for example, transforms from an orthorhombic lattice to monoclinic at 96°C, quickly reverting back on cooling.

More complex semi-regular or aperiodic crystalline systems also occur, in living things (*e.g. below left*) and in quasicrystals, rapidly cooled metal alloys (*e.g. the five-fold Al-Mn system below right*).

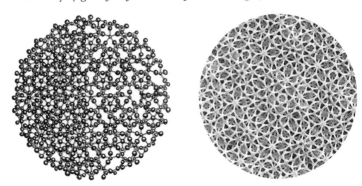

Simple Cubic

Body-Centred Cubic

hexagonal System

Face-Centred Cubic

Simple Monoclinic

Base-Centred Monoclinic

Rhombohedral System

Body-Centred Tetragonal

Simple Tetragonal

Body-Centred Orthorhombic

Simple Orthorhombic

Face-Centred Orthorhombic

Triclinic System

Base-Centred Orthorhombic

15

HYDROGEN AND HELIUM
the first two elements

Hydrogen makes up three quarters of all known matter in the universe and is a large part of most stars. The first element, and the simplest atom, it consists of one proton orbited by one electron.

Hydrogen gas is *diatomic*, which means it is happiest when two atoms covalently bond to form one molecule, H_2. Highly explosive in air, it burns rapidly with oxygen to create water. Under immense pressures and temperatures (everyday conditions in the cores of giant planets like Jupiter and Saturn) hydrogen becomes metallic.

The second element in the periodic table is *helium*. It has two protons, two electrons, two neutrons (99.99% of the time), and is the second most abundant element in the material universe, almost a quarter of it. With two electrons completely filling the 1s-orbital, helium is happy to stay independent and rarely reacts with other elements. It is the first of the *noble* (or *inert*) gases, each of which have full outer electron orbitals. Surprisingly, helium was unknown on Earth until 1870 when it was discovered through *spectrographic analysis* of sunlight, a fingerprinting technique for elements (*lower opposite*). Lighter than air, though twice as heavy as hydrogen, the specks of helium formed here quickly float off into outer space. It is a much safer gas than hydrogen in balloons, and when inhaled produces a squeaky voice.

Beyond its common form, hydrogen has two isotopes, *deuterium*, with one neutron, and *tritium*, with two. Tritium is a rare and unstable beast, decaying into the light helium isotope *helium-3* by changing a neutron into a proton via radioactive beta decay (*see page 36*).

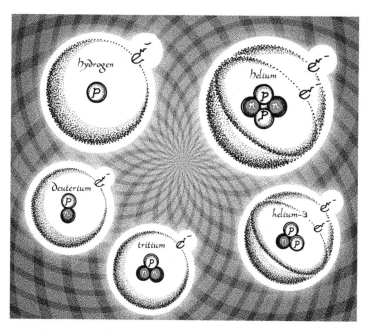

hydrogen and helium: with two isotopes of hydrogen and one of helium showing the number of protons (p), neutrons (n) and electrons (e⁻)

Each element absorbs light in a unique way producing dark bands at specific places across the electromagnetic spectrum: in this way even very distant stars can be analyzed to discover their chemistry.

ALKALI & ALKALINE-EARTH METALS
the violent world of the s-block

The first real group of the periodic table is known as the *alkali metals*, IA (*leftmost column, below*). Soft and silvery-white, they all have a single outer *s*-orbital electron which they enthusiastically lose to form singly charged +1 ions, making them very electropositive.

Lithium, the first member of the group, and the third element, is the lightest metal and floats on water. *Sodium*, immediately below, floats and fizzes as it oxidises, regularly bonding with chlorine to make common salt (NaCl). Next down, *potassium* is the second lightest metal, oxidizing rapidly in air and bursting into flames when wet. Lower still, both *Cæsium*, the most electropositive element, and *rubidium* explode on contact with air. *Francium*, the final member of this vigorous family, is radioactive.

Moving across one column, we meet group IIA, the rare earth metals, *beryllium, magnesium, calcium, strontium, barium* and *radium*. Marginally less electropositive, they gladly form double-charge +2 ions, losing both their outer electrons. They are denser than their group I neighbours, with higher melting and boiling points.

A wire dipped in compounds of these elements will produce characteristic colours when held in a flame. Excited electrons jump between orbitals, losing their energy as little packets of light, *photons*, on the way back down to their normal state (*lower opposite*).

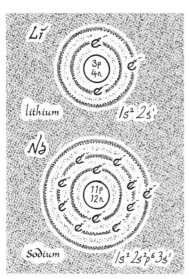

Li

3p
4n

e^- e^- e^-

lithium $\quad 1s^2 2s^1$

Na

11p
12n

e^- ...

Sodium $\quad 1s^2 2s^2 p^6 3s^1$

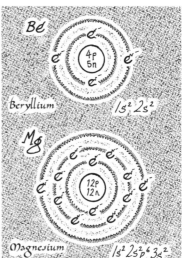

Be

4p
5n

e^- ...

Beryllium $\quad 1s^2 2s^2$

Mg

12p
12n

e^- ...

Magnesium $\quad 1s^2 2s^2 p^6 3s^2$

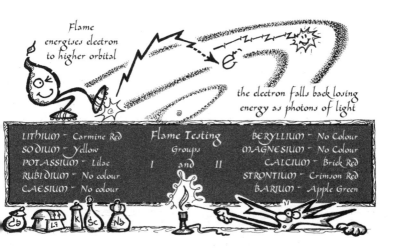

Flame
energises electron
to higher orbital

e^-

the electron falls back losing
energy as photons of light

LITHIUM - Carmine Red   Flame Testing   BERYLLIUM - No Colour
SODIUM - Yellow   Groups   MAGNESIUM - No Colour
POTASSIUM - Lilac   CALCIUM - Brick Red
RUBIDIUM - No colour   I   and   II   STRONTIUM - Crimson Red
CAESIUM - No colour   BARIUM - Apple Green

THE P-BLOCK
metals, metalloids and non-metals

Elements five to ten are the first members of the *p-block*. One to six electrons inhabit three new double-teardrop shaped *p*-orbitals arranged at right angles around the nucleus (*see page 38*). At room temperature and pressure they appear as solids (carbon and aluminium), liquids (bromine) and gases (nitrogen and chlorine), depending on the balance of their interatomic and intermolecular forces. The left-hand side of the *p*-block mostly shines with solid metals. Ductile, malleable and conductive because of their footloose outer electrons, most metals can be stretched into wires, squished into sheets, or combined into alloys.

Crossing the block from left to right, we move from metals to non-metals. These tend to be dull brittle solids, liquids, or gases that are poor conductors of heat and electricity. In this small corner are found many of the players the game of life, such as carbon, oxygen and nitrogen, whose compounds form the backbone of living things and organic chemistry (*opposite top, and page 28*).

In between metals and non-metals lie the *metalloids*, a diagonal streak of ambiguous elements with aspects of both. Among these are the semiconductors, boron, silicon, germanium and arsenic, which form the minds of our computers and electronic gizmos due to their outer valence electrons' ability to jump about inside their nearly full shells.

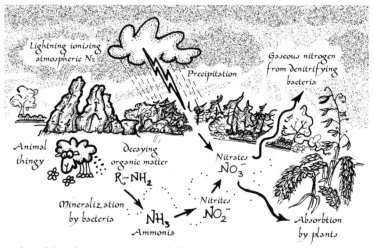

Lightning ionising atmospheric N_2

Precipitation

Gaseous nitrogen from denitrifying bacteria

Animal thingy

Decaying organic matter
$R-NH_2$

Mineralization by bacteria

NH_3
Ammonia

Nitrites
NO_2^-

Nitrates
NO_3^-

Absorbtion by plants

To be useful to plants the strong N_2 triple bond has to be broken by nitrogen fixing bacteria

5 6	6 6	7 7	8 8	9 10
B +3	**C** -3+2+4	**N** -3+1+5	**O** -2	**F** -1
13 14	14 14	15 16	16 16	17 18
Al +3	**Si** -1+2+4	**P** -3+4+5	**S** -2+4+6	**Cl** -1
31 38	32 42	33 42	34 46	35 44
Ga +3	**Ge** +2+4	**As** -3+3+5	**Se** -2+4+6	**Br** -1+1+5
49 66	50 70	51 70	52 78	53 74
In +3	**Sn** +2+4	**Sb** -3+3+5	**Te** -2+4+6	**I** -1+1+5+7
81 124	82 126	83 126	84 125	85 125
Tl +1+3	**Pb** +2+4	**Bi** +3+5	**Po** -2+4	**At** ?
	114 **Uuq**		116 **Uuh**	

Metalloids run diagonally across the p-block marking the changeover between metals and non-metals.

Boron isn't boring
I can tell you that
This bug isn't resting
It's quite dead out flat
All ready for the cooking in
A flameproof pyrex dish
of
borosilicates !!
The fifth atom's
special wish.

BORON IS NOT BORING

CARBON & SILICON
organic and virtual thinking materials

Twenty-three percent of you is carbon. The sixth element underpins *organic* chemistry, the fabric of life, from DNA and proteins in our cells to once living stuff, plastics and fossil fuels. Coming in a dazzling array of molecules, carbon is neither electropositive or negative. A non-metal, it combines with many other elements and also extensively with itself, creating long chains and rings (*see page 55*). Multiple π-bonds smear electrons between atoms to give double and triple bonds.

Carbon arranges itself into several different *allotropes*. In diamond crystals every atom bonds to four others in a hard tetrahedral grid (*opposite top right*) whereas in graphite, a soft crystalline solid found in charcoal and pencils, flat planes of carbon rings slide easily over each other (*opposite top left*). Each atom here joins to three others, the π-bonds enabling it to conduct electricity. Other allotropes include spherical *buckminsterfullerenes*, intriguing *nanotubes* and *graphene*, all of which have amazing structural and conductive properties.

Directly beneath carbon in the periodic table is *silicon*, a metalloid semi-conductor. Carbon life is mirrored by silicon logic in the buzzing microchip mazes of purified silicon crystals, doped with elements like gallium or arsenic to alter their electronic properties.

Stable silicon compounds cover much of the earth as rocks and minerals, such as quartz-rich sand and fine-grained aluminium silicate clays. The Earth's crust consists of 60% aluminium silicate *feldspars* ($K AlSi_3O_8$-$NaAlSi_3O_8$-$CaAl_2Si_2O_8$), followed by quartz (SiO_2), then olivine $(Mg,Fe)_2SiO_4$. Clays have unusual life-mimicking chemistries and perhaps contributed to biological evolution.

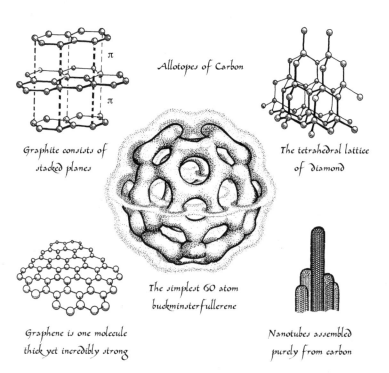

Allotopes of Carbon

Graphite consists of
stacked planes

The tetrahedral lattice
of diamond

The simplest 60 atom
buckminsterfullerene

Graphene is one molecule
thick yet incredibly strong

Nanotubes assembled
purely from carbon

In silicon dioxide (SiO_2) each silicon atom is tetrahedrally bonded to four
oxygen atoms to create quartz crystals and the sand on our beaches.

OXYGEN & SULPHUR
the over and underworlds of group VI A

A fifth of the air we breathe is *oxygen*. After hydrogen and helium, it is the third most abundant element in the universe. Highly reactive, oxygen exerts a strong pull on other atoms to gain the two electrons needed to fill its outer orbitals. Redox (reduction-oxidation) reactions play a fundamental role on our planet and are vital to life, involved in everything from metabolizing food to sending nerve impulses.

On Earth, free oxygen forms an odourless diatomic gas (O_2). High in the atmosphere, naturally occurring triatomic *ozone* (O_3) protects us from harmful ultraviolet radiation, yet lower down its powerful oxidizing effects can damage living organisms. The ten most common compounds in the Earth's crust are *oxides*; just under half is sand (silicon dioxide SiO_2), a third is magnesium oxide (MgO) and much of the rest is rock salt, iron(II) oxide (FeO). Water (H_2O) is another essential oxygen compound, covering 71% of our world's surface.

Below oxygen in the periodic table, so mirroring its chemistry in many ways, is the smelly underworld of *sulphur*. Usually a brittle, pale yellow solid, it has a profusion of multi-atom ring and chain allotropes, burning in air to create sulphur dioxide (SO_2). Combining with water in clouds, it becomes tree-scolding sulphurous acid rain. Sulphur is less electronegative than oxygen, and hydrogen sulfide (H_2S) acts differently than water, hydrogen bonding having little influence. To us a fiendishly toxic gas with a rotten egg smell, colonies of creatures nevertheless live in the dark on energy metabolised by sulphur-breathing bacteria beside deep ocean volcanic vents bubbling hydrogen sulfide. Indeed, life on Earth may have started down in hot wet places like these.

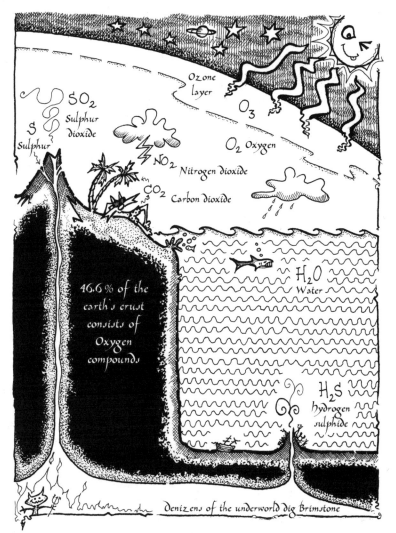

Ozone layer

O_3

SO_2
Sulphur dioxide

S
Sulphur

O_2 Oxygen

NO_2 Nitrogen dioxide

CO_2 Carbon dioxide

46.6% of the earth's crust consists of Oxygen compounds

H_2O
Water

H_2S
hydrogen sulphide

Denizens of the underworld dig Brimstone

WATER & ACIDS
making a splash

Water is the most common molecule in the universe. One oxygen and two hydrogen atoms, H_2O is two thirds of all of us.

The water molecule is *polarised*. The pull from the oxygen atom gives the hydrogen atoms a slight positive charge (*opposite top left*) resulting in extensive networks of hydrogen-bonded molecules creating sixfold snow crystals (*opposite*), surface tension and the fluctuating crystal lattice we drink (*below*). Water's attraction allows it to unsettle and dissolve even its own molecules into ions, one water molecule donating a proton (H^+) to another, forming a solution of hydronium (H_3O^+) and hydroxide (OH^-) ions.

Acids are compounds that actively donate protons when in solution, attacking metals to liberate hydrogen gas. Gladly accepting the protons, *bases* are compounds which are soapy and bitter, combining with an acid to form a *salt* and water. An everyday example is hydrochloric acid and sodium hydroxide combining into common salt and water: $HCl + NaOH = NaCl + H_2O$ Certain metal oxides, hydroxides, amines and the group I and II alkalis are particularly caustic in this respect. Other *Lewis* acids and bases respectively accept or donate an electron pair subject to the solvents used, and not all acids need water to do their corrosive work.

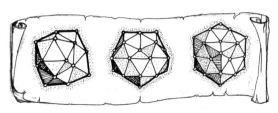

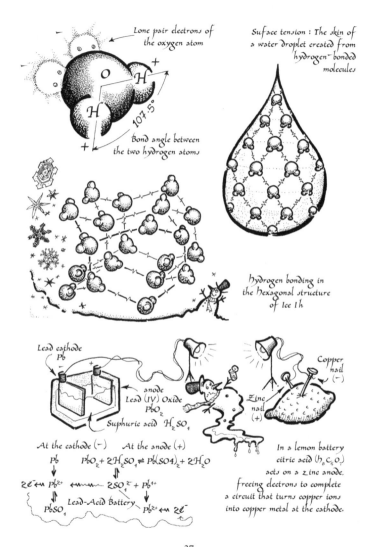

Lone pair electrons of
the oxygen atom

Surface tension: The skin of
a water droplet created from
hydrogen-bonded
molecules

O

H

H

$107.5°$

Bond angle between
the two hydrogen atoms

Hydrogen bonding in
the Hexagonal structure
of Ice Ih

Lead cathode
Pb

anode
Lead (IV) Oxide
PbO_2

Suphuric acid H_2SO_4

Copper
nail
(−)

Zinc
nail
(+)

At the cathode (−)   At the anode (+)

Pb $PbO_2 + 2H_2SO_4 \rightleftharpoons Pb(SO4)_2 + 2H_2O$

$2e^- \leftarrow Pb^{2+}$ $2SO_4^{2-} + Pb^{4+}$

$PbSO_4$ Lead-Acid Battery $Pb^{2+} \leftarrow 2e^-$

In a lemon battery
citric acid $(h_8C_6O_7)$
acts on a zinc anode.
freeing electrons to complete
a circuit that turns copper ions
into copper metal at the cathode.

ORGANIC CHEMISTRY
biocosmic molecules

Carbon creates literally millions of compounds and is vital to life as we know it. *Hydrocarbons* are made only of carbon and hydrogen and form the basis of the oils that power industry. Crude oil, a mix of hydrocarbons, is separated out into fractions through a heated column (*opposite top right*), in which the lighter oils with fewer carbon atoms rise the highest. Organic chemistry deals with combinations of carbon and hydrogen, often with oxygen and nitrogen, sulphur, phosphorus and others.

Easily linking into lattices and sharing bonds gives carbon the ability to *polymerise*, joining simple *monomer* molecules into long chains of repeating units (*opposite top left*). Polymers are the foundations of the fantastic plastics that are used everywhere from satellites to false teeth.

The building blocks of organic chemistry are the *functional groups* which define how compounds behave (*see page 55*). Methanol (CH_3OH), the first alcohol (-OH group), is poisonous whilst ethanol (C_2H_5OH), the second, can liven up an evening. Change the alcohol into an aldehyde (-CHO), however, and it can ruin the next morning (*lower opposite*).

Carbon molecules occur in vast quantities in space, drifting in giant molecular clouds that form the nurseries of new stars. Natural reactions turn simple molecules such as carbon monoxide (CO), water (H_2O), ammonia (NH_3), methane (CH_4), cyanic acid (HOCN) and formaldehyde (H_2CO) into more complex ones like acetone (($CH_3)_2CO$), ethyl alcohol (CH_3CH_2OH) and cyanodecapentayne ($HC_{10}CN$). Even some fullerenes (e.g. C_{70}) float free. Hydrocarbons are abundant on other planets, raining down to form lakes on Saturn's moon Titan. Interstellar laboratories often create the essential elements of life, so perhaps we are not alone.

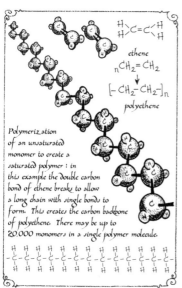

$$H\!>\!C=C\!<\!H$$

ethene

$$nCH_2 = CH_2$$

$$[-CH_2-CH_2-]_n$$

polyethene

Polymerization of an unsaturated monomer to create a saturated polymer: in this example the double carbon bond of ethene breaks to allow a long chain with single bonds to form. This creates the carbon backbone of polyethene. There may be up to 20,000 monomers in a single polymer molecule.

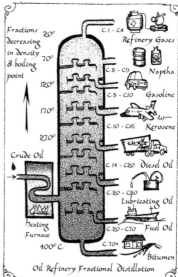

Fractions decreasing in density & boiling point

20°   C.1 - C.4   Refinery Gases

70°   C.5 - C.9   Naptha

120° C.5 - C.10 Gasoline

170° C.10 - C.16 Kerosene

270° C.14 - C.20 Diesel Oil

Crude Oil

C.20 - C.50 Lubricating Oil

C.20 - C.70 Fuel Oil

Heating Furnace

400°C C.70+ Bitumen

Oil Refinery Fractional Distillation

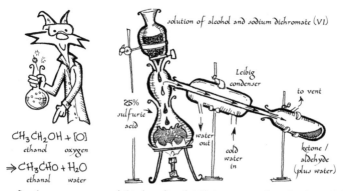

solution of alcohol and sodium dichromate (VI)

Leibig condenser

25% sulfuric acid

to vent

water out

cold water in

ketone / aldehyde (plus water)

$$CH_3CH_2OH + [O]$$
ethanol   oxygen

$$\Rightarrow CH_3CHO + H_2O$$
ethanal   water

Recipe for a hangover: preparing an aldehyde/ketone from alcohol by heating under reflux. The dichromate (VI) is reduced by the alcohol to a chromium (III) ion, whilst the alcohol is oxidized to an aldehyde or ketone.

HALOGENS & THE NOBLE GASES
ups and downs at period's end

The universe's most vigorous and inert elements are found in the final two columns of the periodic table. The members of group VIIA, the *halogens*, are just one electron short of a full shell, and aggressively form compounds to complete it. All elements, except helium, neon and argon, bond with a halogen to form a *halide*.

The ninth element, and easily the most electronegative, is *fluorine*, a pale green-yellow diatomic gas which combines fanatically with almost anything, attacking compounds to form *fluorides*. The rest of the group are also intensely reactive, particularly that rascal chlorine, which is why it is so good at killing bugs in bleaches (*see opposite*).

With one more proton and one more electron added, we finally meet the quiet, solipsistic group VIIIA. With all the slots of their electron orbitals full, they are closed to business and on the whole content not to react with anything. That said, *xenon* does form (with effort) a few compounds with feisty fluorine and neighbouring oxygen, and a few helium and krypton compounds also exist, so the former name of this group, the *inert gases*, was changed in the 1960s to the slightly less lazy *noble gases*.

When you next see glowing neon signs, picture those full orbitals frantically buzzing with jumping electrons.

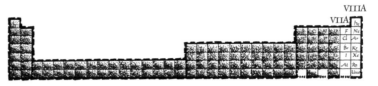

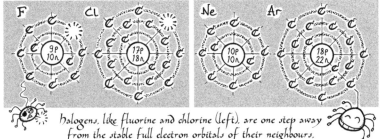

halogens, like fluorine and chlorine (left), are one step away
from the stable full electron orbitals of their neighbours,
the noble gases neon and argon (right).

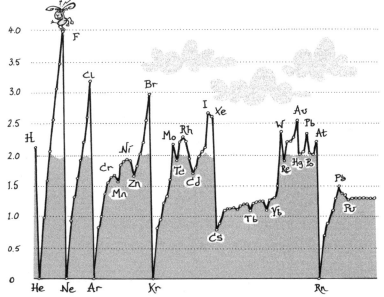

Electronegativity measures how easily an atom will attract electrons in a molecule:
the very reactive halogens occupy the highest peaks whilst the noble gases
(with the notable exception of xenon) sleep quietly in the deepest valleys.

THE TRANSITION METALS
gold, silver, copper and iron in the d-block

The next zone we encounter on our travels across the periodic kingdom is a series of metals starting at scandium where the first of ten electrons begins filling the set of *3d* orbitals *inside* the *4s (see page 9)*. Most members of this series heartily lose one or more electrons to form a bewildering array of brightly coloured compounds.

The *transition metals* are hard and strong, their similar structures allowing them to be mixed into useful *alloys*. Copper and zinc combine into brass, and mercury, the only metal liquid at room temperature, forms alloys called *amalgams*. With a dash of carbon, iron creates steel, becoming even harder with an added splash of vanadium, molybdenum, or chromium. Iron's magnetic attraction is due to the unbalanced magnetic moments of unpaired electrons in its outer *d*-orbital. Several near neighbours, notably nickel, cobalt and manganese, also exhibit varying degrees of paramagnetism.

Titanium has a reputation for both strength and corrosive resistance, and is thus ideally suited for flying machines and rocket ships.

The enduring popularity of shiny gold, silver and copper is in part due to their dependable stability. Excellent conductors of electrons and heat, they have many applications in electronics and optics, also looking very pretty in rings, crowns, coins and other baubles.

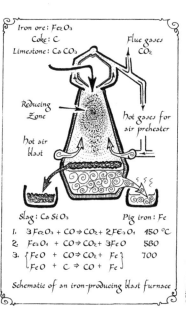

Iron ore: Fe_2O_3
Coke: C
Limestone: $CaCO_3$

Flue gases CO_2

Reducing Zone

hot gases for air preheater

hot air blast

Slag: $CaSiO_3$ Pig iron: Fe

1. $3Fe_2O_3 + CO \Rightarrow CO_2 + 2Fe_3O_4$ 450 °C
2. $Fe_3O_4 + CO \Rightarrow CO_2 + 3FeO$ 580
3. $\begin{cases} FeO + CO \Rightarrow CO_2 + Fe \\ FeO + C \Rightarrow CO + Fe \end{cases}$ 700

Schematic of an iron-producing blast furnace

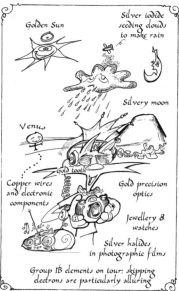

Golden Sun

Silver iodide seeding clouds to make rain

Silvery moon

Venus

Gold tooth

Copper wires and electronic components

Gold precision optics

Jewellery & watches

Silver halides in photographic films

Group 1B elements on tour: skipping electrons are particularly alluring

Z	Symbol	outer d- and s-electrons	oxidation states
21	Sc	1-2	+3
22	Ti	2-2	+2+3+4
23	V	3-2	+2+3+4+5
24	Cr	5-1	+2+3+6
25	Mn	5-2	+2+3+4+7
26	Fe	6-2	+2+3
27	Co	7-2	+2+3
28	Ni	8-2	+2+3
29	Cu	10-1	+1+2
30	Zn	10-2	+2
39	Y	1-2	+3
40	Zr	2-2	+4
41	Nb	4-1	+3+5
42	Mo	5-1	+6
43	Tc	5-2	+4+6+7
44	Ru	7-1	+3
45	Rh	8-1	+3
46	Pd	10-0	+2+4
47	Ag	10-1	+1
48	Cd	10-2	+2
57	La	1-2	+3
72	Hf	2-2	+4
73	Ta	3-2	+5
74	W	4-2	+6
75	Re	5-2	+4+6+7
76	Os	6-2	+3+4
77	Ir	7-2	+3+4
78	Pt	9-1	+2+4
79	Au	10-1	+1+3
80	Hg	10-2	+1+2
89	Ac	1-2	+3
104	Rf	?	+4
105	Db	?	?
106	Sg	?	?
107	Bh	?	?
108	Hs	?	?
109	Mt	?	?
110	Uun	?	?
111	Uuu	?	?
112	Uub	?	?

The d block transition metals, showing for each element the number of protons (top left), the number of electrons in the outer d- and s-orbitals (top right), and oxidation states (below). Electrons skip from outer s to d-orbitals when the latter are half-full or full.

THE F-BLOCK & SUPERHEAVIES
enormous atoms and islands of stability

At lanthanum, element fifty-seven, a *5d* orbital starts to fill before something strange happens; the next electron drops into a previously hidden *4f* orbital inside the full *6s*, *5s* and *5p* orbital sets, taking the electron from the *5d* with it. The *5d* orbitals wait patiently until the *4fs* are full, apart from one halfway hiccup at gadolinium, where an electron briefly flickers up to the *5d*.

Quietly spreading from lanthanum to lutetium the *lanthanides*, or *rare-earth* metals, fill a fourteen place set of *4f* orbitals. Overshadowed by their *5s* and *5p* sets, only subtle chemical differences are found in this series.

Below the lanthanides, the radioactive *actinides* play much the same trick, as two electrons begin a *6d* orbital only to quit the job and turn within to fill a *5f* orbital instead. Uranium is the last natural element; artificially made atoms now fill a seventh of the periodic table.

Beyond the *f*-block at rutherfordium, a fourth transition series starts filling a *6d* orbital. These superheavy elements tend to be highly radioactive and unstable due to uhappy ratios of protons to neutrons (*see map opposite*). Elements with up to 118 protons have been fleetingly created in particle accelerators. Around elements *114p* 'eka-lead' or *184n* there may possibly be rare islands of stability where a few isotopes with balanced nuclei have significant lifetimes of minutes rather than seconds.

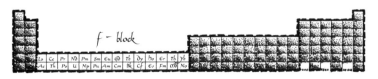

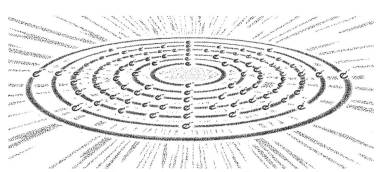

The electrons of Plutonium (not the order they fill in - see page 7):
$1s^2\ 2s^2\ 2p^6\ 3s^2\ 3p^6\ 3d^{10}\ 4s^2\ 4p^6\ 4d^{10}\ 4f^{14}\ 5s^2\ 5p^6\ 5d^{10}\ 5f^6\ 6s^2\ 6p^6\ 7s^2$

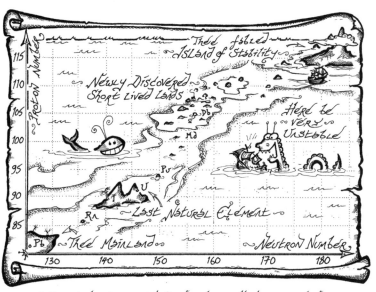

A Mappe to show ye wayye to lands afar whence stable elementes might forme
wythe a balancing of neutron and protonne numbers

RADIOACTIVITY
nuclear fizzicks

Held together by immensely strong forces, an atomic nucleus contains huge amounts of energy. Unstable nuclei rebalance by spitting out radioactive emissions as protons and neutrons join (*fusion*), or split off (*fission*). An isotope's radioactivity halves over its *half-life*; the less stable it is the faster it *decays*. Uranium-238 has a half-life of 4.5 billion years, yet with ten neutrons less, uranium-228 halves every fifth of a second!

All living things are slightly radioactive, constantly absorbing carbon-14 and tritium generated by cosmic rays. At death, we stop gathering these isotopes, and archæologists use the 5,730-year half-life of carbon-14 to date historically interesting goo.

Beyond bismuth, all elements have *radioisotopes* that undergo α-*(alpha)* decay, the nucleus expelling an α-*particle* (a helium nucleus, *see page 17*). Thin clothing should prevent α-particles from ionizing the unwary. Excess neutrons in a nucleus cause β-*(beta)* decay, where a neutron converts into a proton, releasing a speeding electron (a β-*particle*); protective apparel or 2 mm of aluminium halts these beasties. Often found alongside α- or β-decays, γ-*(gamma) rays* are high-energy photons that carry off energy as electromagnetic radiation, requiring a good few inches of lead to prevent them from zipping through you.

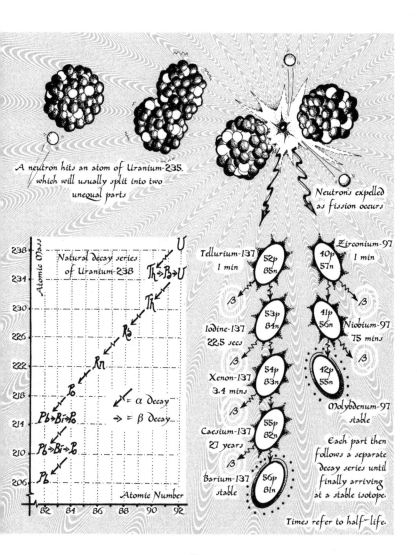

A neutron hits an atom of Uranium-235, which will usually split into two unequal parts

Neutrons expelled as fission occurs

Natural decay series of Uranium-238

Atomic Mass

238 — U

234 — Th → Pa → U

230 — Th

226 — Ra

222 — Rn

218 — Po

214 — Pb → Bi → Po

210 — Pb → Bi → Po

206 — Pb

↖ = α decay

→ = β decay

Atomic Number

82   84   86   88   90   92

Tellurium-137   52p 85n   1 min

β

Iodine-137   53p 84n   22.5 secs

β

Xenon-137   54p 83n   3.4 mins

β

Caesium-137   55p 82n   27 years

β

Barium-137   56p 81n   stable

Zirconium-97   40p 57n   1 min

β

Niobium-97   41p 56n   75 mins

β

Molybdenum-97   42p 55n   stable

Each part then follows a separate decay series until finally arriving at a stable isotope.

Times refer to half-life.

37

ORBITAL STRUCTURES
the whirly world of the very small

At the scale of fundamental particles like the electron, energy comes in discrete packets, or *quanta*. Strangely, everything down here behaves like both particles *and/or* waves, depending on your perspective.

If an atom is pumped up with energy, excited electrons whizzing around the nucleus make sudden *quantum* energy jumps into new orbital levels that fit together like a buzzing ethereal flower.

Mathematical *wave functions* can predict the probability of finding an electron in a specific place. Most of time the electron will be within the main part of its density plot (*opposite left*). However there is always a slim chance it could be somewhere else entirely.

Each orbital can be inhabited by two electrons, which need to have opposite spins to each other. The sphere (*opposite top right*) represents the primary *1s* orbital and its twins may be anywhere, including in the nucleus. The second orbital set, the *2p*, fills as shown below. Three double teardrop shapes reflect around a nuclear *nodal plane* where electrons hardly ever go. As further orbitals fill, new electrons are forced into more and more exotic dances (*lower opposite, and page 58*).

Weirdly, the wavy nature of electrons means that we can't know their position *and* speed at the same time. Just looking at something so tiny radically alters its behaviour. A small but measurable *uncertainty* creeps in.

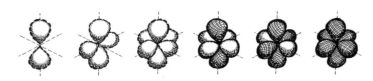

38

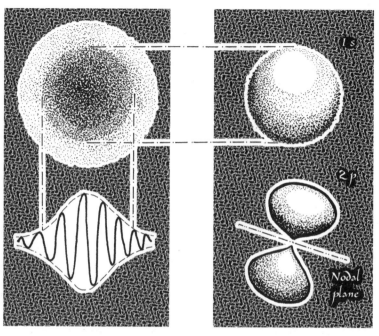

Finding the elusive electron : from wave packet to atomic orbital

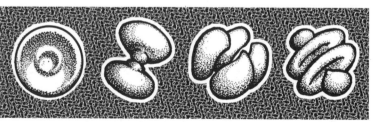

2s   3p   4d   5f

An assortment of orbital shapes : note that only s orbitals lack a nodal plane

Materials Science
the future is invisible

Peering into atoms and molecules to understand how they work has enabled the manipulation of matter on finer and finer scales, producing immensely strong alloys for deep ocean and space exploration, 3D printers to assemble things a slice at a time, and fuel-cells and solar panels to keep things going when the oil runs out. Theory can predict new chemistry, new possibilities and practical offspring (*opposite top*). Novel plastics, polymers, ceramics and composites help make stuff that is ever more flexible, adhesive, superconducting and heat resistant.

The children of the quantum revolution can be found in the flashing LEDs of everyday electronics, and the silicon microchips of their brains, all of this due to the erratic way electrons wander through certain crystals. Spin engineering, which uses fundamental properties of atoms and electrons, has allowed us to peek harmlessly into compounds or living tissue by nuclear magnetic resonance. Spintronics will probably form the minds of the next generation of computers (*centre opposite*).

Today, we can actually push single atoms about, leading to the science of *nanotechnology* where the odd effects of the quantum realms are harnessed. The size of a nanometer is one billionth of a meter (the length fingernails grow in a second), and applications for nanotech are already present in medicine, electronics and chemical synthesis (*lower opposite*). One day designer molecules may build themselves, and microscopic intravenous nanobioelectronic robots could rid us of disease, or alas turn all they touch into grey goo. Molecular engineering has huge consequences, and with our environment already stressed, we need to be extremely careful about our powerful new toys.

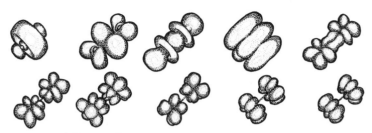

Quintuple Diuranium bonding - using quantum chemistry it has been discovered that two uranium atoms may form molecules held together with five covalent bonds

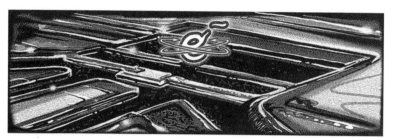

A qubit - at the heart of a quantum computer each qubit is able to hold and measure electrons in a superposition of 0 AND 1 simultaneously. Classical computers can only deal with 0 OR 1

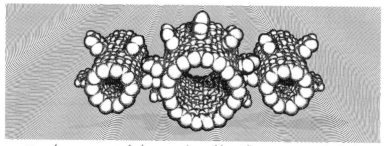

Nanotech gearing - nanotechnology opens the possibilities of engineering at a molecular level

QUIRKY QUARKS
and curious quantum effects

Deep inside the *nucleons* (protons and neutrons) lurks an even smaller realm. Energy and matter down here have such a close relationship that it's sometimes difficult to tell them apart.

Both nucleons are made of three *quarks*, particle-like matter fields that are the building blocks of the everyday universe. The quarks we usually encounter are called *up* and *down*. Peeking inside a proton (*opposite top left*) we find two up quarks and one down quark. Up quarks have an electric charge of $+\frac{2}{3}$, whilst downs carry $-\frac{1}{3}$. Adding the three quark charges gives the proton's total of +1. The neutron is composed of a slightly different crew—two down quarks and one up quark, which cancel out, leaving it electrically neutral (*opposite top right*).

Binding quarks together is the special job of the *nuclear strong force*. Intriguingly, instead of the two charges of electricity, the strong force has three charges to balance which can be likened to the three primary colours of light. Mixing red, green and blue light together produces a neutral white, and similarly, each nucleon carries three different 'colour' charged quarks which combine into a neutral overall charge. The colour force is carried by peculiar fundamental particles called gluons.

Weird quantum effects rule this deep level of matter. When quarks are pulled apart, pairs of quarks and mirror-image antiquarks seeimgly magically pop into being. Since the strength of the strong force remains constant over its short range, regardless of distance, like unbreakable elastic bands stretched to their limit, it becomes more efficient to borrow energy from the quantum vacuum than to overcome the gluon bonds that hold the quarks together (*lower opposite*).

The Proton
Two Up quarks and one Down
$2/3 + 2/3 - 1/3 = +1$

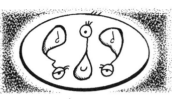

The Neutron
One Up and two Down
$2/3 - 1/3 - 1/3 = 0$

Quarks exchanging gluons to balance
the red, green, and blue colour charges
of the strong force, overall giving a
neutral charge across the nucleon.

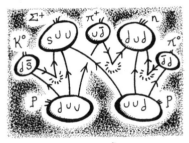

A Feynman diagram of two protons (p)
transforming into a neutron (n), a strange
sigma (Σ^+) particle and a host of mesons from
spontaneous quark/antiquark creation

Confinement is due to the strong force colour charge remaining constant as distance
increases: as quarks are pulled apart, it is energetically easier for quark/antiquark
pairs to be created than to stretch the gluon fields holding them together.
Only very very short lived top quarks have been observed alone

THE FOUR FORCES
holding the universe together

Everything in the universe interacts with everything else through four universal forces, carried by four types of wave-particles called *gauge bosons*.

The *electromagnetic* force is carried by *photons*. Light, x-rays, radio waves and microwaves are all different frequency wrigglings of electromagnetic fields. This dual-aspect force attracts electrons to protons and causes most annihilations between matter and antimatter (*see below*). It is also veritably the prime mover of all chemical reactions.

Acting over distances the size of a nucleus, the *strong* force binds quarks by exchanging eight types of *gluon*, the carriers of colour charge. The strong force only affects quarks and gluons.

Particle decay is governed by the *weak* force, which acts over extremely short ranges. Changing a down quark into an up one, for example, transforms a neutron into a proton and hence an element into the next in the periodic table. Carried by *W* and *Z vector bosons*, the weak force also allows neutrinos, very light fundamental particles, to perform their rare interactions with everyday matter.

Gravity, although by far the weakest force, nevertheless operates over almost infinite distances, extending its grip over all matter. Finding a quantum explanation for gravity has had its up and downs, though a carrier boson, the *graviton*, has been tentatively described.

Positron (Anti-electron)

Electron

$E = Mc^2$

Photons

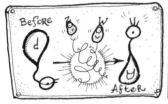

The electromagnetic force can operate over large distances: it shapes electron orbits and hence controls chemical behaviour in atoms

The strong force acts over distances the size of a nucleus, shaping and holding the quarks within together

The weak force has a very short range indeed and is responsible for quark transformation and neutrino interactions

Gravity acts across very great distances and connects all types of matter, from galaxies to the atoms of a cat

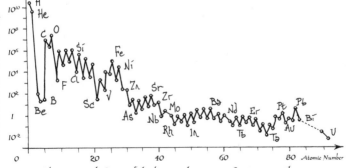

Relative cosmic abundances of the elements: the zig-zag is due to atoms with even numbers of protons and electrons being slightly more stable than ones with odd numbers

QUARKS, LEPTONS & MESONS
the fundamental families of matter

Everyday matter is made of up and down quarks, and two *lepton* cousins, the electron and neutrino, in a first generation of indivisible fundamental particles (*opposite top*). At higher energies, like cosmic rays, a second, heavier, family of four is found, *charm* and *strange* quarks, with the *muon* and *muon–neutrino*. At extreme energies a third, even more massive, family appears. Each member also has a mirror-image antimatter twin.

All matter has *spin*, a quantum version of angular momentum. Quarks and leptons have spin $\frac{1}{2}$, needing to 'turn' around twice to look the same (*see slugs below*). Particles with half spin= ($\frac{1}{2}, \frac{3}{2}, \frac{5}{2}$ …) are called *fermions*, bound by the *Pauli exclusion principle*, which forbids two fermions to be in the same quantum state. For example, electron pairs need to have opposite (up or down) spin to share an atomic orbital. *Baryons*, composites of three quarks, like protons and neutrons, are also fermions.

Particles with integer spin (0, 1, 2…) are called *bosons*, the force-carrying gauge bosons and *mesons* (quark/antiquark pairs) being examples (*see appendix page 54*). Bosons can sit in the same quantum state as each other, like photons in a laser beam acting in tandem.

Under certain conditions, even numbers of fermions can act like bosons, giving rise to weird quantum behaviours on a macroscopic scale, like ultra-runny superfluids with no viscosity, or superconductive electrons joining in Cooper pairs to flow without resistance.

.............BOSONS.............
The SPIN 2 mirror slug has half-turn symmetry

The SPIN 1 tumble snail rotates a full circle

FERMIONS
The SPIN 1/2 moebius slug twists through two circles 720°

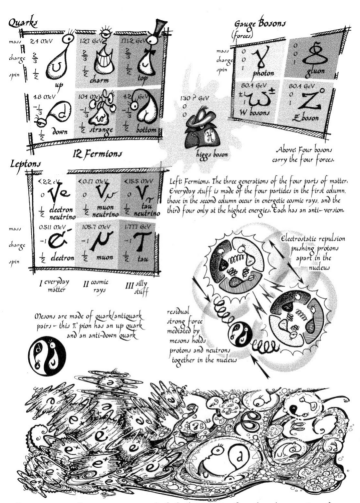

Quarks

mass	2.4 MeV	1.27 GeV	171.2 GeV
charge	⅔	⅔	⅔
spin	½	½	½
	up	charm	top

mass	4.8 MeV	104 MeV	4.2 GeV
charge	-⅓	-⅓	-⅓
spin	½	½	½
	down	strange	bottom

12 Fermions

130 ? GeV
0
0
higgs boson

Gauge Bosons
(forces)

mass	0	0
charge	0	0
spin	1	1
	photon	gluon

mass	80.4 GeV	80.4 GeV
charge	±1	0
spin	1	1
	W bosons	Z boson

*Above: Four bosons
carry the four forces.*

Leptons

	<2.2 eV	<0.17 MeV	<15.5 MeV
	0	0	0
	½	½	½
	electron neutrino	muon neutrino	tau neutrino

mass	0.511 MeV	105.7 MeV	1.777 GeV
charge	-1	-1	-1
spin	½	½	½
	electron	muon	tau

I everyday matter	II cosmic rays	III silly stuff

*Left: Fermions. The three generations of the four parts of matter.
Everyday stuff is made of the four particles in the first column,
those in the second column occur in energetic cosmic rays, and the
third four only at the highest energies. Each has an anti-version.*

*Electrostatic repulsion
pushing protons
apart in the
nucleus*

*Mesons are made of quark/antiquark
pairs - this π pion has an up quark
and an anti-down quark*

*residual
strong force
mediated by
mesons holds
protons and neutrons
together in the nucleus*

*The Quantum Vacuum: there is no empty space. At the Planck scale a seething sea of virtual particles pop in and out of existence.
An electron surrounded by virtual counterparts (left). Quark/antiquark mesons created by balancing energy debts (right).*

EXOTIC PARTICLES
and sub-atomic siblings

Energetic cosmic rays from outer space, high-speed charged particles like protons and helium nuclei, with a few electrons and tiny amounts of antimatter, constantly hit our planet, ionizing atoms in the upper atmosphere to create cascades of subatomic particles. Visible as the glowing polar auroras, branching streamers of *hadrons* (quark composites of baryons and mesons) shed electrons, positrons and γ-rays to cause further chain reactions, short-lived exotic particles rapidly decaying into more stable, lower energy ones (*opposite top*).

   In experiments re-creating the extreme energies of the early universe, accelerators collide protons and other matter at near light speed to reveal hidden structures. Detectors trace the curling paths of hundreds of colliding and transmuting particles (*below*). Some are higher energy versions, or *resonances*, of others, linked together by symmetrical patterns into families. One, called the *eightfold way* (after a Buddhist doctrine), charts family relationships as octets (*lower opposite*) correlating charges, spins and other characteristics. The quantum orchestra plays some surprising tunes as nature reveals her subtle harmonics.

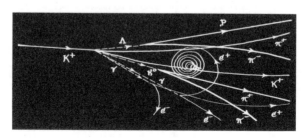

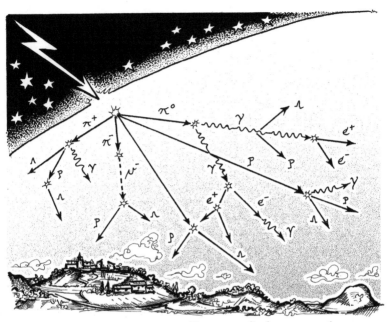

A high energy cosmic ray sets off a cascade of sub-atomic particles

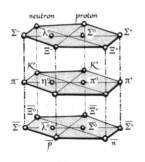

Eight fold way octets

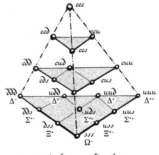

Spin 3/2 baryon family tree

49

QUANTUM THEORIES
interpretations and entanglements

To this day no one understands *why* the application of matrix mechanics in the complex plane predicts electron orbitals, nor *how* an entangled pair of photons seem to be able to communicate instantly regardless of how far apart they are separated (*below*), nor *why*, in the twin slit experiment, a single fired photon, atom, or molecule interferes as a *wave* (so passes through both slits) when unobserved, yet behaves as a *particle* (going through just one slit) when observed (*opposite top*).

There are different *formulations* of quantum mechanics. All work beautifully, making precise predictions for building our mobile phones and computers, but each leads to a different *interpretation* of what is happening, and so the true nature of reality. The most widely taught is the *Copenhagen interpretation*, which holds that microscopic reality is *actually* created by observation—*looking* causes the collapse of the probabilistic wave function. Next up is the *Many Worlds interpretation*, which says that the world constantly divides, so the moment the cat looks in the box he really does create one world in which the scientist is alive and another one in which he is dead (*lowest opposite*). Another, the *Transactional interpretation* allows the future to affect the past, while the *Bohm interpretation* sees the entire universe as a single entangled whole. There are many more. The truth is that we do not know, but we do know that the future is quantum, and that it's going to be different.

The Twin Slit experiment:
Wave or particle? Unobserved,
single photons passing through
two slits produce an interference
pattern, like waves would. however
if we try to see which slit the photon
went through, the pattern dissapears
as if the photons were discrete particles

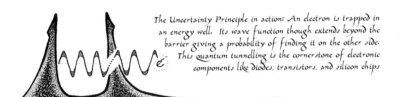

The Uncertainty Principle in action: An electron is trapped in
an energy well. Its wave function though extends beyond the
barrier giving a probability of finding it on the other side.
This quantum tunnelling is the cornerstone of electronic
components like diodes, transistors, and silicon chips

The Casimir Effect: Very close parallel
conducting plates are pushed together
by fluctuations in the quantum vacuum.
Since only electromagnetic waves with
resonant wavelengths can fit in the
cavity, the field pressure between the
plates is lower inside than that outside

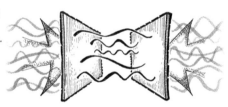

Kat and Schrödinger: A scientist is in a closed
box with a quantum system that gives a 50/50
chance of survival. The probability wave function
says that the system exists in a superposition of
states until observed where the hapless subject is
both alive and dead. Only when measured will
the wave function seem to decohere to one state.

STRINGS AND THINGS
bubbles and branes

The more we probe into the wispy knottings of wave-matter, the more the colossal energies bound up in the subatomic realm present a field day for adventurous mathematicians. Many attempts to find a *Theory of Everything* invoke additional dimensions to the usual four of spacetime. Quantum models that include gravity need eleven or so dimensions to work, and use *supersymmetries* to relate fundamental families and forces to phenomena. Most envisage an initial singularity connected through a single unified force, folded and condensed into the familiar four.

The knitting of *superstrings*, hypothetical one-dimensional standing wave threads which loop and resonate, along with the lure of *supergravity* are different views over a much larger vista, the mysterious M-theory. This pictures the universe as ripples in an infinitely vast and thin *p-brane*, a membrane spanning many dimensions stretching through hyperspace. Was the Big Bang branes colliding, our little universe being merely the interference patterns where they crossed? Others theories suggest that we are just one universe amongst a multitude, or possibly a 3D projection cast from a 2D information horizon.

Elegant geometries perhaps underlie the inner workings of reality, with particles, spins and forces all surfing around a deeper symmetry. It could be that any description of nature will of necessity remain incomplete, the subtleties of the quantum realm obscuring attempts to peer too far into nature's infinitely layered onion. Formed from hidden harmonies, could we be only as real as the holographic sparkles in a sunny pool, or the fractal rainbow swirls on the surface of a soap bubble?

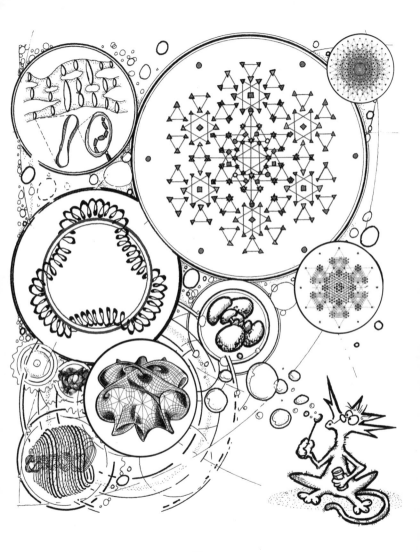

CONSTANTS AND HADRONS

Electron Mass	m_e	$9.1091534 \times 10^{-31}$ kg
		= 0.5110 MeV
Electron charge	e	1.602189×10^{-19} C
		= 4.8030×10^{-10} esu
Proton mass	m_p	1.672648×10^{-27} kg
		= 1836.1 x electron mass
Neutron mass	m_n	$1.6749545 \times 10^{-27}$ kg
Atomic mass unit	u	1.66054×10^{-27} kg
Avogadro's no.	N_A	6.022045×10^{23} mol^{-1}
Bohr radius	a_o	$0.52917706 \times 10^{-10}$ mol^{-1}
Boltzmann const.	k	1.380662×10^{-23} J K^{-1}
Faraday constant	F	9.648456×10^{4} C mol^{-1}
Gas constant	R	8.31441 J K^{-1} mol^{-1}
H_2O triple point	T_{tpw}	273.16 K
Ice point temp.	T_{ice}	273.1500 K
Mol. vol. gas	V_M^o	2.241383×10^{-2} m^3 mol^{-1}
Perm. of vacuum	μ_o	$4\pi \times 10^{-7}$ H m^{-1}
Permittivity const.	ε_o	8.8542×10^{-12} F m^{-1}
Planck constant	h	6.626176×10^{23} mol^{-1}
Planck length	l_p	1.616×10^{-35} m
Planck time	t_p	5.319×10^{-44} s
Planck mass	m_P	2.1777×10^{-8} kg
Rcp. fine str. const.	$1/\alpha$	137.036
Rydberg const.	R_H	1.097373×10^{7} m^{-1}
Speed of light	c	2.99792458×10^{8} m s^{-1}
1 angström (Å)		10×10^{-10} m
1 atmosphere		101325 N m^{-2} (Pa)
1 calorie (cal)		4.184 joules (J)
1 celsius (°C)		273.150 Kelvin (K)
°celsius		5/9 (°fahrenheit - 32)
1 curie (Ci)		3.7×10^{10} s^{-1}
1 erg		2.390×10^{-11} kcal
1 esu		3.3356×10^{-10} C
1 eV		1.60218×10^{-19} J
1 eV/molecule		96.485 kJ mol^{-1}
1 kcal mol^{-1}		349.76 cm^{-1}, 0.0433 eV
1 kJ mol^{-1}		83.54 cm^{-1}
1 wave no. (cm^{-1})		2.8591×10^{-3} kcal mol^{-1}

Baryons	Symbol	Mass	Quarks	Charge	Spin
Proton	N$^+$	938	uud	+1	1/2
Neutron	N^o	940	ddu	0	1/2
Sigma$^+$	Σ$^+$	1198	uus	+1	1/2
Sigma^o	Σ^o	1192	dus	0	1/2
Sigma-	S-	1197	dds	-1	1/2
Lambda^o	Λ^o	1116	dus	0	1/2
Xi^o	Ξ^o	1315	uss	0	1/2
Xi$^-$	Ξ$^-$	1321	dss	-1	1/2
Sigma$^+$	Σ$^+$	938	uus	+1	1/2
Delta^{++}	Δ^{++}	1231	uuu	+2	3/2
Delta$^+$	Δ$^+$	1232	duu	+1	3/2
Delta^o	Δ^o	1234	ddu	0	3/2
Delta$^-$	Δ$^-$	1235	ddd	-1	3/2
Sigma^{*+}	Σ^{*+}	1189	uus	+1	3/2
Sigma^{*o}	Σ^{*o}	1193	dus	0	3/2
Sigma^{*-}	Σ^{*-}	1197	dds	-1	3/2
Xi^{*o}	Ξ^{*o}	1315	uss	0	3/2
Xi^{*-}	Ξ^{*-}	1321	dss	-1	3/2
Omega$^-$	Ω$^-$	1672	sss	-1	3/2

Mesons

Pi$^+$	π$^+$	140	ud̄	+1	0
Pi^o	π^o	135	uū, dd̄	0	0
Pi$^-$	π$^+$	140	dū	-1	0
Eta^o	η^o	547	uū, dd̄, ss̄	0	0
Eta prime	η'	958	uū, dd̄, ss̄	0	0
Kaon$^+$	K$^+$	494	us̄	+1	0
Kaon^o	K^o	498	ds̄	0	0
Rho$^+$	ρ$^+$	770	ud̄	+1	1
Rho^o	ρ^o	770	uū, dd̄	0	1
Omega	ω	782	uū, dd̄	0	1
Phi	φ	1020	ss̄	0	1
K^{*+}	K^{*+}	892	us̄	+1	1
K^{*o}	K^{*o}	892	ds̄	0	1
J/ψ	ψ	3097	cc̄	0	1
Upsilon	Y	9460	bb̄	0	1

Note: Over 200 baryons and 36 mesons are known

CARBON CHEMISTRY

Name	Functional Group	Suffix	First Member Formula	Name	General Formula
Alkane	$R-CH_3$	-ane	$H-\overset{H}{\underset{H}{C}}-H$	methane	C_nH_{2n+2}
Alkene	$\overset{H}{\underset{R}{}}C=C\overset{H}{\underset{R}{}}$	-ene	$\overset{H}{\underset{H}{}}C=C\overset{H}{\underset{H}{}}$	ethene	C_nH_{2n}
Alkyne	$R-C\equiv C-R'$	-yne	$H-C\equiv C-H$	ethyne	C_nH_{2n-2}
Alcohol	$R-\overset{OH}{\underset{H}{C}}-R'$	-anol	$H-\overset{OH}{\underset{H}{C}}-H$	methanol	$C_nH_{2n+1}OH$
Aldehyde	$R-\overset{O}{C}-H$	-anal	$H-\overset{O}{C}-H$	methanal	$C_nH_{2n-1}OH$
Ketone	$R-\underset{O}{C}-CH_3$	-anone	$H-\underset{O}{C}-CH_3$	methanone	$C_nH_{2n}O$
Carboxylic acid	$R-\underset{O}{C}-OH$	-anoic acid	$H-\underset{O}{C}-OH$	methanoic (formic) acid	
Amine	H_2N-R	-anamine	$CH_3-N\overset{H}{\underset{H}{}}$	methylamine	
Ester	$R-\overset{O}{C}-OR'$	-ate	$CH_3-C\overset{O}{\underset{OCH_3}{}}$	carboxylic acid + alcohol	
Amide	$R-\overset{O}{C}-NHR'$	-ide	$CH_3-\overset{O}{C}-NHR'$	carboxylic acid + amine	
Ether	$R-O-R'$	-oxy... -ane	CH_3-O-CH_3	methoxy methane	
Cyclopropane	$\overset{H}{\underset{H}{}}C\overset{H}{\underset{H}{C}}\overset{H}{\underset{H}{}}C$	Benzene	(benzene ring)	or ⬡ or ⬡	

Branched alkanes:
branch name changes from -ane to -yl.

H_2C-CH_3 ethyl branch

$H_3C-\underset{CH_2}{}-CH_2-\underset{CH}{}-CH_2-CH_3$
　　　6　　5　　4　　3　　2　　1
3-ethyl hexane

CH_3 methyl branch

$H_3C-CH_2-\underset{CH}{}-CH_3$
　5　　4　　3　　2　　1
2-methyl pentane

Prefix (no. of carbon atoms): Meth-1, Eth-2, Prop-3, But-4, Pent-5, Hex-6, Hept-7, Oct-8, Non-9, Dec-10

The Periodic Table

Group	IA 1	IIA 2		IIIB 3	IVB 4	VB 5	VIB 6	VIIB 7	VIII 8
Period 1	H (0) · 1 · Hydrogen · 1.00079 · 1310								
Period 2	Li (B 3) · 3 · Lithium · 6.941 · 519	Be (H 4) · 4 · Beryllium · 9.01218 · 900							
Period 3	Na (B 11) · 11 · Sodium · 22.9878 · 494	Mg (H 12) · 12 · Magnesium · 24.3050 · 736							
Period 4	K (B 20) · 19 · Potassium · 39.0983 · 418	Ca (C 20) · 20 · Calcium · 40.0785 · 590		Sc (H 24) · 21 · Scandium · 44.9559 · 632	Ti (H 26) · 22 · Titanium · 47.867 · 661	V (B 28) · 23 · Vanadium · 50.9415 · 648	Cr (B 24) · 24 · Chromium · 51.9961 · 653	Mn (C 25) · 25 · Manganese · 54.9380 · 716	Fe (B 28) · 26 · Iron · 55.845 · 762
Period 5	Rb (B 48) · 37 · Rubidium · 85.4678 · 402	Sr (F 50) · 38 · Strontium · 87.62 · 548		Y (H 50) · 39 · Yttrium · 88.9059 · 616	Zr (H 50) · 40 · Zirconium · 91.224 · 669	Nb (B 52) · 41 · Niobium · 92.9064 · 653	Mo (B 56) · 42 · Molybdenum · 95.94 · 694	Tc (H 55) · 43 · Technetium · 97.9072 · 699	Ru (H 58) · 44 · Ruthenium · 101.07 · 724
Period 6	Cs (B 78) · 55 · Caesium · 132.905 · 376	Ba (B 82) · 56 · Barium · 137.327 · 502	57 – 70 Lanthanide series *	Lu (H 104) · 71 · Lutetium · 174.967 · 481	Hf (H 104) · 72 · Hafnium · 178.49 · 531	Ta (B 108) · 73 · Tantalum · 180.948 · 760	W (B 110) · 74 · Tungsten · 183.84 · 770	Re (H 112) · 75 · Rhenium · 186.207 · 762	Os (H 116) · 76 · Osmium · 190.23 · 841
Period 7	Fr (H 136) · 87 · Francium · 223.02 · 381	Ra (H 138) · 88 · Radium · 226.025 · 510	89 – 102 Actinide series **	Lr (? 157) · 103 · Lawrencium · 262.110 · 444	Rf (? 153) · 104 · Rutherfordium · 263.113 · 490	Db (? 157) · 105 · Dubnium · 262.114 · ?	Sg (? 157) · 106 · Seaborgium · 266.122 · ?	Bh (? 157) · 107 · Bohrium · 264.125 · 740	Hs (? 161) · 108 · Hassium · 269.134 · ?

* Lanthanide series

La (H 82) · 57 · Lanthanum · 138.906 · 540	Ce (C 82) · 58 · Cerium · 140.116 · 665	Pr (H 82) · 59 · Praseodymium · 140.908 · 556	Nd (H 82) · 60 · Neodymium · 144.24 · 607	Pm (H 84) · 61 · Promethium · 144.913 · 556	Sm (R 90) · 62 · Samarium · 150.36 · 544	Eu (B 90) · 63 · Europium · 151.964 · 548	Gd (H 94) · 64 · Gadolinium · 157.25 · 594	Tb (H 94) · 65 · Terbium · 158.925 · 648

** Actinide series

Ac (C 138) · 89 · Actinium · 227.028 · 669	Th (C 142) · 90 · Thorium · 232.038 · 674	Pa (T 140) · 91 · Proactinium · 231.0356 · 568	U (O 146) · 92 · Uranium · 238.029 · 385	Np (O 144) · 93 · Neptunium · 237.048 · 604	Pu (M 150) · 94 · Plutonium · 244.064 · 585	Am (H 148) · 95 · Americium · 243.061 · 578	Cm (H 151) · 96 · Curium · 247.070 · 581	Bk (H 94) · 97 · Berkelium · 247.070 · 601

IUPAC interim naming system for new elements : o-nil-(n), 1-un-(u), 2-bi-(b), 3-tri-(t), 4-quad-(q), 5-pent-(p), 6-hex-(h), 7-sept-(s), 8-oct-(o), 9-enn-(e)

VIII	VIII	IB	IIB	IIIA	IVA	VA	VIA	VIIA	VIIIA or O
9	10	11	12	13	14	15	16	17	18

OF THE ELEMENTS

He — H, 2, Helium, 4.0026, 2370

| | | | | B (R, 6, 5) | C (H, 6, 6) | N (H, 7, 7) | O (M, 8, 8) | F (M, 10, 9) | Ne (C, 10, 10) |

Element	Crystal Structure	Neutrons	Atomic No.	Name	Atomic Weight	First Ionization Energy
He	H	2	2	Helium	4.0026	2370
B	R	6	5	Boron	10.811	799
C	H	6	6	Carbon	12.0107	1090
N	H	7	7	Nitrogen	14.0067	1400
O	M	8	8	Oxygen	15.9994	1310
F	M	10	9	Fluorine	18.9984	1680
Ne	C	10	10	Neon	20.1797	2080
Al	C	14	13	Aluminium	26.9815	577
Si	C	14	14	Silicon	28.0855	786
P	TC	16	15	Phosphorus	30.9738	1060
S	O	16	16	Sulphur	32.066	1000
Cl	O	18	17	Chlorine	35.4527	1260
Ar	C	22	18	Argon	39.948	1520
Co	H	32	27	Cobalt	58.9332	757
Ni	C	30	28	Nickel	58.6934	736
Cu	C	34	29	Copper	63.546	745
Zn	H	34	30	Zinc	65.39	908
Ga	O	38	31	Gallium	69.723	577
Ge	C	42	32	Germanium	72.61	762
As	R	42	33	Arsenic	74.9216	966
Se	H	46	34	Selenium	78.96	941
Br	O	44	35	Bromine	79.904	1140
Kr	C	48	36	Krypton	83.80	1350
Rh	C	58	45	Rhodium	102.906	745
Pd	C	60	46	Palladium	106.42	803
Ag	C	60	47	Silver	107.868	732
Cd	H	66	48	Cadmium	112.411	866
In	T	66	49	Indium	114.818	556
Sn	T	70	50	Tin	117.710	707
Sb	R	70	51	Antimony	121.760	833
Te	H	78	52	Tellurium	127.60	870
I	O	74	53	Iodine	126.904	1010
Xe	C	78	54	Xenon	131.29	1170
Ir	C	116	77	Iridium	192.217	887
Pt	C	117	78	Platinum	195.078	866
Au	C	118	79	Gold	196.967	891
Hg	R	122	80	Mercury	200.59	1010
Tl	H	124	81	Thallium	204.383	590
Pb	C	126	82	Lead	207.2	716
Bi	M	126	83	Bismuth	208.980	703
Po	C	126	84	Polonium	208.982	812
At	?	125	85	Astatine	209.987	920
Rn	C	136	86	Radon	222.017	1040
Mt	?	159	109	Meitnerium	268.139	?
Ds	?	162	110	Darmstadtium	272.146	?
Rg	?	161	111	Roentgenium	272.154	?
Cn	?	165	112	Copernicium	277	?
Uut	?	173	113	Ununtrium	286	?
Uuq	?	175	114	Ununquadium	289	?
Uup	?	174	115	Ununpentium	289	?
Uuh	?	177	116	Ununhexium	293	?
Uus	?	177	117	Ununseptium	294	?
Uuo	?	176	118	Ununoctium	294	?

Element	Crystal Structure	Neutrons	Atomic No.	Name	Atomic Weight	First Ionization Energy
Dy	H	98	66	Dysprosium	162.50	657
Ho	H	98	67	Holmium	164.930	?
Er	H	98	68	Erbium	167.26	?
Tm	H	98	69	Thulium	168.934	?
Yb	C	104	70	Ytterbium	173.04	598
Cf	HC	153	98	Californium	251.080	608
Es	C	158	99	Einsteinium	252.083	619
Fm	?	157	100	Fermium	257.095	627
Md	?	157	101	Mendelevium	258.098	635
No	?	157	102	Nobelium	259.101	642

Legend:

Crystal Structure (See below for key) — Number of neutrons (most abundant or stable isotope)

ATOMIC NUMBER

Chemical Symbol

Name of Element

Atomic Weight (Average relative mass) — First Ionization Energy (kJ mol⁻¹)

B body centred cubic C cubic close packing H hexagonal close packing M monoclinic O orthorhombic R rhombohedral (trigonal) T tetragonal TC triclinic

EXAMPLES OF ELECTRON ORBITALS

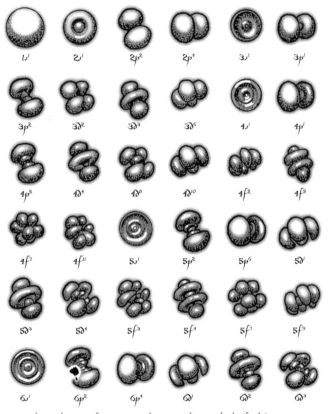

1s¹ 2s¹ 2p² 2p⁴ 3s¹ 3p¹

3p² 3d² 3d³ 3d⁶ 4s¹ 4p¹

4p⁸ 4d¹ 4d⁸ 4d¹⁰ 4f² 4f⁵

4f⁷ 4f¹¹ 5s¹ 5p² 5p⁶ 5d¹

5d³ 5d⁴ 5f³ 5f⁴ 5f⁷ 5f⁹

6s¹ 6p² 6p⁴ 6d¹ 6d² 6d³

These enclosure surfaces represent the various electron orbitals of a hydrogen atom. Each traces the quantum probability wavefunction of an electron pair as they juggle with the many attractive and repulsive forces inside the atom to give a fantastically labyrinthine interweave. S orbitals are spherical, p orbitals have twin lobes, d orbitals are four-fold or enjoy a single donut, whilst double donuts and six lobed shapes belong to the f orbitals.

58